Kucherov Olexandr

Power of Knowledge in Economy and Nature

Kucherov Olexandr

Power of Knowledge in Economy and Nature

Knowledge is power, which creates the organized motion and never implements any work

LAP LAMBERT Academic Publishing

Impressum / Imprint

Bibliografische Information der Deutschen Nationalbibliothek: Die Deutsche Nationalbibliothek verzeichnet diese Publikation in der Deutschen Nationalbibliografie; detaillierte bibliografische Daten sind im Internet über http://dnb.d-nb.de abrufbar.
Alle in diesem Buch genannten Marken und Produktnamen unterliegen warenzeichen-, marken- oder patentrechtlichem Schutz bzw. sind Warenzeichen oder eingetragene Warenzeichen der jeweiligen Inhaber. Die Wiedergabe von Marken, Produktnamen, Gebrauchsnamen, Handelsnamen, Warenbezeichnungen u.s.w. in diesem Werk berechtigt auch ohne besondere Kennzeichnung nicht zu der Annahme, dass solche Namen im Sinne der Warenzeichen- und Markenschutzgesetzgebung als frei zu betrachten wären und daher von jedermann benutzt werden dürften.

Bibliographic information published by the Deutsche Nationalbibliothek: The Deutsche Nationalbibliothek lists this publication in the Deutsche Nationalbibliografie; detailed bibliographic data are available in the Internet at http://dnb.d-nb.de.
Any brand names and product names mentioned in this book are subject to trademark, brand or patent protection and are trademarks or registered trademarks of their respective holders. The use of brand names, product names, common names, trade names, product descriptions etc. even without a particular marking in this works is in no way to be construed to mean that such names may be regarded as unrestricted in respect of trademark and brand protection legislation and could thus be used by anyone.

Coverbild / Cover image: www.ingimage.com

Verlag / Publisher:
LAP LAMBERT Academic Publishing
ist ein Imprint der / is a trademark of
OmniScriptum GmbH & Co. KG
Heinrich-Böcking-Str. 6-8, 66121 Saarbrücken, Deutschland / Germany
Email: info@lap-publishing.com

Herstellung: siehe letzte Seite /
Printed at: see last page
ISBN: 978-3-659-63406-2

Olexandr Kucherov

POWER OF KNOWLEDGE IN ECONOMY AND NATURE

2014

ABSTRACT

To quote Engels about dialectical materialism: "In the present work dialectics is conceived as the science of the most general laws of all motion. Therein is included that their laws must be equally valid for motion in nature and human history and for the motion of thought."

In the book, the pivotal concept within Marxist theory, which has determined the working power and the ruling power in economy, is offered to extend on nature. Power of knowledge appears as a result of their united action and starts implementation a number of the enigmatic phenomena.

It is shown that power of knowledge is a mechanism converting the ocean water heat into the hurricane with his stormy wind, enormous clouds, lightnings and thunder. It is a mechanism converting cold ice of the comet nucleus into its tail, forming hot, brightly luminous plasma that fills the half of sky. And, finally, you will know how to use the power of knowledge in your business.

It illustrates how mechanism with power of knowledge can really work in you business.

Contents

Introduction

In the manuscript, the class nature of society is offered to extend on all nature, by analogy with the pivotal concept within Marxist theory, which has determined the working class and the ruling class, in accordance with their specific functions they must execute in a reproduction. The Marxist class concept is identical, both for ideas of society and for ideas of nature. Thus, the unique property of Lorenz power to accomplish the ruling but not executing work allows us to formulate the known phrase "knowledge is power" as a law of nature: knowledge is power, which rules by motion and never implements any work.

The physical knowledge power, or Lorenz power, is able to organize motion of magnetic dipoles and to create an informative reality. In the manuscript the informative reality is got from the Ampere equalization by the field theory methods. This phenomenon strictly submits to the primarily laws of dialectical materialism (transformation of quantity into quality, and vice versa; interpenetration of opposites; and negation of the negation).

For this reason, examining Lorenz power as power that organizes motion in some set direction, many difficult phenomena of nature become clear. It is shown that the informative reality gives power to the hurricanes/cyclones, comets and astrophysical masers. This statement is confirmed in a number of facts. The results of visible, infrared, and radio spectrometry are written done by informative reality theory. It describes adequately, in details polarization and the masses spectral dates.

The implementation of knowledge power in science gives substantial results. For example, it improves functional possibilities and increases an efficiency of the system that prepares engine's air – fuel mixture.

Chapter 1:
The power of knowledge

The independent existence of informative reality, which submits to its own laws, had been formulated by a Ukrainian scientist Ermoshenko M.M. in his fundamental work "Information in the system of productions relations" [11]. In this connection, we will consider the informative reality from a point of the class structure of society. It has often been remarked that class is perhaps the pivotal concept within Marxist theory, and Marx himself provided a systematic definition of class. One chapter in Capital is devoted to analysis of class. There are also numerous passages elsewhere in Capital, and in other works, where Marx presents many of the elements of a rigorous definition of class.

1.1. Ruling material force of society

In recent years, as part of the general attempt by Marxist theorists to clarify rigorously the conceptual foundations of Marxism, there has been considerable effort placed on developing the concept of class.

The concept is to consider those wage earners who fall outside the working class not as a segment of the petty bourgeoisie but rather as a new class in its own right, called "Professional and Managerial Class" (PMC). This class is defined by the specific function it plays in the reproduction of class relations [12,32].

Managers and supervisors occupy an opposing location between the working class and the capitalist class. Like the working class they are excluded from control over money capital (that is from basic decisions about allocation of investments and the direction of accumulation), but unlike workers they have a certain degree of control of the physical means of production and over the labor of workers within production [7].

Scientific ideas were class-situated for Marx. With Engels, he had written decisively about the role of ruling-class [28]:

"The ideas of the ruling class are in every epoch the ruling ideas; i.e. the class which is the ruling material force of society, is at the same time its ruling intellectual force. The class which has the material means of production at its disposal, has control at the same time over the means of mental production, so that thereby, generally speaking, the ideas of those who lack the means of mental production are subject to it. The ruling ideas are nothing more than the ideal expression of the dominant material relationships... grasped as ideas; hence of the relationships which make the one class the ruling one, therefore, the ideas of its dominance."

It is clear from the above said that society can be divided into a working class, which executes work, and the class of capitalists, top and supervisor managers, who have a power of knowledge what kind of work they need to execute.

Practical social mediation applies to all knowledge; this is the Marxist ideas, and in particular the principles of dialectical materialism. These principles are identical, both for ideas of society and for ideas of nature. "One can look at history from two sides and divide it into the history of nature and the history of men. The two sides are, however, inseparable; the history of nature and the history of men are dependent on each other so long as men exist" [6].

The purpose of the manuscript is to find the analogy of the ruling force in nature.

1.2. Ruling force in nature

The scalar multiply of force vector $\vec{F}$ on the vector of motion $d\vec{r}$ determines work A:

$$\vec{F} \bullet d\vec{r} = A. \tag{1}$$

Where $\vec{F}$ is the working force, which executes work, $A>0$.

But if the force vector $\vec{F}$ is perpendicular to motion $d\vec{r}$, thereby, scalar multiply equals zero:

$$\vec{F} \bullet d\vec{r} = 0. \tag{2}$$

Where $\vec{F}$ is the ruling force, it executes no work, $A=0$.

It is often alleged that Lorenz force $\vec{F}$ always operates across direction of motion $d\vec{r}$ and, consequently, it does no work on moving charges. The question appears: if the force never executes any work, so what does it execute? It is possible to give such a working answer: this is a ruling material force of nature.

1.3. The law of informative reality

The phrase *scientia est potentia* is a Latin aphorism often claimed to mean "knowledge is power". It is commonly attributed to Sir Francis Bacon in Novum Organum [49].

Thus, property of Lorenz force $\vec{F}$ to accomplish the ruling, and not execute work allows the author to use the known utterance "knowledge is power" to formulate the law of informative reality: knowledge is power, which creates the organized motion and never implements any work [16].

The numerous researches show that the action of law of informative reality is inherent not only to economic development, but also it spreads to all nature. In such understanding the law of informative reality explains development of matter that specifies a transition from the mechanical understanding of physic to the dialectical understanding of physic, where motion of bodies is directed by the knowledge power.

In accordance with the laws of dialectical materialism, informative reality carries out development of nature.

To quote Engels about dialectical materialism [6]: "In the present work dialectics is conceived as the science of the most general laws of all motion. Therein is included that their laws must be equally valid for motion in nature and human history and for the motion of thought." Engels took Hegel's logic as the most suggestive result of classical philosophical investigations. Engels himself believed that scientific investigators abstract dialectical laws (for the greater part unknowingly) from "the history of nature and human society." He agreed with Hegel, that there are primarily three such laws: transformation of quantity into quality, and vice versa; interpenetration of opposites; and negation of the negation.

It is popularly believed that the mechanical understanding of the world negates the development of nature and supposes a heat death.

The heat death opened by William Thomson [39] in 1851 is the term that describes the eventual state of any thermodynamics close system, and the universe in particular. Thus, no directed exchange by energy the observed not will, as all types of energy will pass to thermal. Thermodynamics examines the system that is in the state of heat death, as a system in which thermodynamics entropy maximal.

8

William Thomson [39] once wrote: "I believe the tendency in the material world is for motion to become diffused, and that as a whole the reverse of concentration is gradually going on – I believe that no physical action can ever restore the heat emitted from the Sun, and that this source is not inexhaustible; also the motions of the Earth and other planets are losing vis viva which is converted into heat; and that although some vis viva may be restored for instance to the earth by heat received from the sun or by other means, the loss cannot be precisely compensated and I think that it is probably under compensated.

Compensation would require a creative act or an act possessing similar power."

According to L. D. Landau [26], the rotation of entropy is necessary to be searched in area of general relativity theory: as the Universe is the system that is in the variable gravitational field, the increase entropy law is not applicable here.

There is an opposite paradigm of natural and technical sciences, which was worked out by Boyle during 50th of XVII of century [13]. Boyle, the action of the Universe creation took place once and forever, God created natural laws started a mechanism in an action and nature became a legal object from the machines of natural philosophy. That is the rotation of entropy in nature is stopped up initially.

We will study the creative force of nature that creates the informative reality.

Chapter 2:
The theory
of informative reality

It is possible to assert that informative reality exists in nature similar to objective reality. Objective reality is given in feelings but informative reality is given in rules which explain the certain way of matter movement.

We will give the definition of informative reality through its difference from objective reality, namely through physical work.

2.1. Ruling force of the informative reality

In this connection, we will consider the informative reality created by Lorenz force. This theory describes the fact of the presence of magnetic properties in revolving water molecules. The rotation of positive protons creates a solenoid, as shown on the figure 1. Direction of the magnetic field is determined by the rule of gimlet and magnetic induction B is created by Lorenz force:

$$B = \frac{mV}{qR},\qquad (3)$$

where m is the total mass of two protons, q is a charge of two protons, V is the linear speed of protons, and R is the radius of rotation.

Thus, the equation notes the fact of the presence of magnetic properties in revolving water molecules.

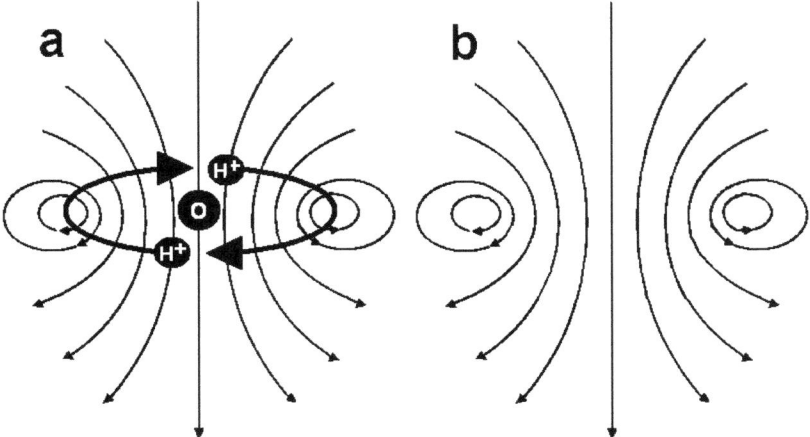

Figure 1. Solenoid. a) The motion of positively charged atoms of hydrogen around the negatively charged atom of oxygen creates the magnetic field. b) The informative reality.

The figure 1 shows the informative reality. The rotation of positive protons in the water molecule creates it. The informative reality is created by one rotary magnetic molecule. Many rotary magnetic molecules co-operate with each other and create collective informative reality with great number additional properties. This state is possible owing to the fact that integral equalization of Ampere the methods of the field theory.

2.2. Condition of parallelism

It is possible to calculate the force $\vec{G}$ which operates on the solenoid in the external magnetic field with magnetic induction vector $\vec{B}$ from the integrated Ampere law:

$$\vec{G} = J \oint_l d\vec{l} \times \vec{B} . \tag{4}$$

We will use the generalized Stoks theorem to compute the integrated Ampere law:

$$\oint_l d\vec{l} \times \vec{B} = \int_S (d\vec{S} \times \nabla) \times \vec{B} , \tag{5}$$

where the operator ∇ is generally accepted.

On condition that the linear size of solenoid $\vec{p}_m$ is compared to distance, on which intensity external magnetic field changes substantially, we can replace the work of current on the area by the dipole moment of a solenoid:

$$\vec{G} = (\vec{p}_m \times \nabla) \times \vec{B} . \tag{6}$$

We will apply the known identity of vector analysis:

$$(\vec{p} \times \nabla) \times \vec{B} =$$
$$(\vec{p}\nabla) \bullet \vec{B} + \vec{p} \times rot \ \vec{B} - \vec{p} \bullet div \ \vec{B} . \tag{7}$$

Assuming that in the point of disposition of solenoid $\vec{p}_m$ there are no current $\vec{j} = 0$ and, accordingly, such condition implements $rot \ \vec{B} = 0$.

Absence of current is the main difference of the informative reality theory from the theory of plasma physics and theory MGD dynamo, where, vice versa, the current of charged ions plays the basic role.

The statement about complete absence of current and, accordingly, the charged ions could seem strange. In fact, hurricanes/cyclones are full of the electric charged clouds and lightnings. However, charges and lightnings will be considered later, and now we must assume complete absence of current.

Taking into account that $div\ \vec{B} = 0$, force $\vec{G}$ can be defined by the following equation:

$$\vec{G} = (\vec{p}_m \nabla) \bullet \vec{B}, \tag{8}$$

the detailed description can be found in the textbook [30]. The equation shows that resulting force $\vec{G}$, which operates on a small solenoid in the external magnetic field, exists only in gradient of magnetic induction vector.

The equation is similar to the force which operates on a small electric dipole $\vec{p}_e$ in the external electric field. The equation shows that solenoid's energy in the external magnetic field is minimal on condition that the vector of solenoid is oriented along the magnetic field, and maximal on condition that the magnetic moment vector of solenoid is oriented against the magnetic field. Solenoid takes position with minimum energy for the relaxation time. As the magnetic field is stable during any hour, the solenoids have the time for relaxation. This process is stable in time and in space. As a result, four vectors became parallel in each point:

$$\vec{G} \parallel \vec{V} \parallel \vec{p}_m \parallel \vec{B}. \tag{9}$$

Therefore, we got a parallelism [16] of the force $\vec{G}$, speed $\vec{V}$, solenoid $\vec{p}_m$ and magnetic induction vector $\vec{B}$ for every point (each point without exception) of the informative reality field, which is shown on figure 2.

Solenoids cluster along the lines of magnetic field and strengthen the magnetic field as a result of the parallelism. We will notice that property to strengthen the field have solenoids only, but electric dipoles, on the contrary, decrease the external magnetic field.

2.3. Potential of the informative reality

The condition of parallelism allows action of vector operator ∇ in expression for force $\vec{G}$ to be replaced by the gradient from the scalar field B :

$$\vec{G} \;=\; p_{\,m} \, gradB \quad . \tag{10}$$

A type of the equation allows us to consider force $\vec{G}$ as a new field, which is neither magnetic nor electric field. This field arises due to the gradient of magnetic field. We will define the tension $\vec{W}$ of gradient magnetic force in accordance with the rule of vector analysis [20]:

$$\vec{W} \;=\; gradB \quad . \tag{11}$$

The equation gives the nature of the field $\vec{W}$. As it is generally known from the vector analysis, tension $\vec{W}$ of force is potential because it is received as the gradient of a scalar field. Consequently, vector force of the magnetic induction $\vec{B}$ in the potential field $\vec{W}$ changes into a scalar potential B . Moreover, the scalar potential B in the informative reality theory is not force as it takes place in wires with a current for a solenoid. The scalar potential B in the informative reality theory is energy.

The field $\vec{W}$ was received from the condition of stability of informative reality process in time. Stability of this process in the space creates an Informative Reality System. We will find it.

Chapter 3
The Informative Reality System

Motion in a tube of an ideal gas is well-known, which is limited to surfaces, such as pipe, cone and nozzle.

3.1. Tractrices field

The informative reality field was found by Kucherov O.P., Pazdriy Y.E. [19]. The vector of magnetic induction in any point of field is a function in the cylindrical co-ordinates $B(r, \varphi, z)$. Then we can write down equalization of the field theory for magnetic field lines as follows:

$$\frac{dr}{Br} = \frac{dz}{Bz} = \frac{d\varphi}{B_\varphi} \qquad (12)$$

Assuming the condition of the field symmetry in relation to an axis z, $B(\varphi)=const$, the magnetic induction $B(r,z)$ depends only on two parameters:

$$\frac{dr}{Br} = \frac{dz}{Bz} \qquad (13)$$

Solenoid that moves at a speed $\vec{V}$ during time t will overcome distance F. Its projection along a radius F is equal to

r, and along to the rotational axis $-\sqrt{F^2 - r^2}$. Owing to the condition of parallelism, the triangle with the sides of B, Br and Bz is similar to the triangle with the sides of F, r and $\sqrt{F^2 - r^2}$. It is possible to write down for Br and Bz:

$$Br = B\frac{r}{F}; \quad Bz = B\frac{\sqrt{F^2 - r^2}}{F}. \tag{14}$$

Substituting Br and Bz in equalization of the field theory, we obtain the relation between dr and dz:

$$\frac{dr}{r} = \frac{dz}{\sqrt{F^2 - r^2}}. \tag{15}$$

Thus, we got differential presentation of tractrix, which was discovered by Claude Perrault in 1670. After integration we will get the tractrix in an obvious kind:

$$z(r) = F \ln\frac{F + \sqrt{F^2 - r^2}}{r} - \sqrt{F^2 - r^2}, \tag{16}$$

where F is a distance from any point on tractrix to the axis z along tangent (on definition), and r is a cylindrical co-ordinate.

Therefore, magnetic field lines in the informative reality process are tractrices with permanent negative curvature $F(r, \varphi, z) = const$. Thus, the tractrices field creates the Informative Reality System.

3.2. Three elements of the informative reality system

The figure 2 shows the Informative Reality System. It consists of the pumping space, inversion layer, and pushing space.

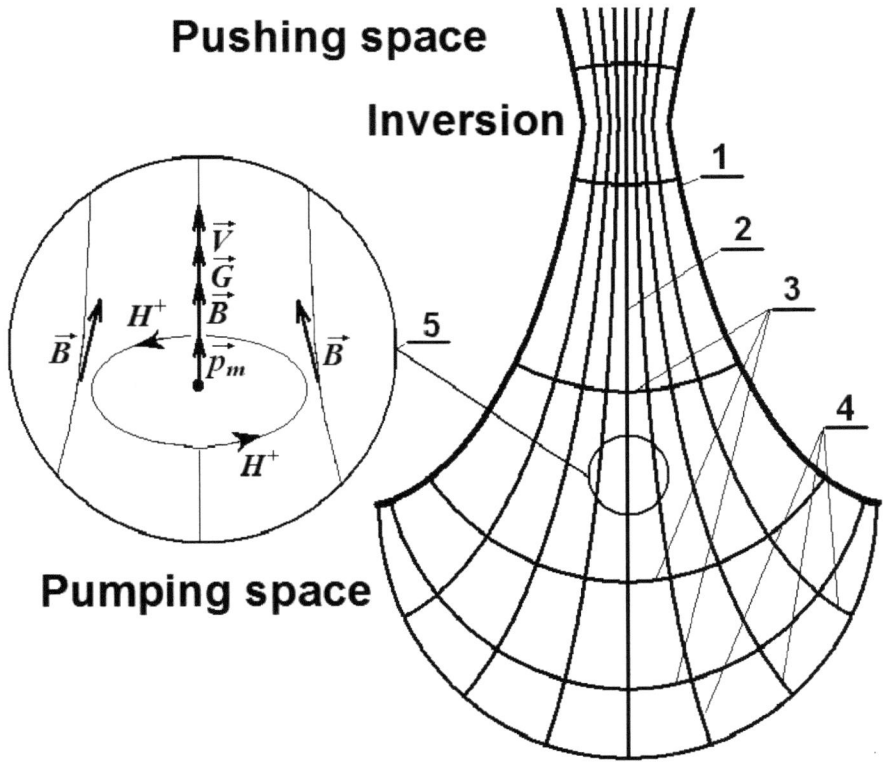

Figure 2. The Informative Reality System.

The field formed by the tractrix rotation around axis z is shown in figure 2. The pumping space consists of the following parts: 1) the tractric nozzle that borders the System; 2) the rotational axis; 3) equal potential surfaces, which are spheres of the permanent positive radius F; 4) magnetic field lines, which are tractrices of the permanent negative radius F; 5) force $\vec{G}$, speed $\vec{V}$, solenoid $\vec{p}_m$, and magnetic induction $\vec{B}$, which are parallel in each point according to equation(9).

The geometrical structure of tractrices field represented by figure 2 and consisting of tractic nozzle 1 is the tractrix rotation around the axis 2. Magnetic field lines 4 are tractrices of the radius F, which move along the rotational axis. Surfaces of identical potential 3 are spheres of equal radius F, which move along the rotational axis. How we can see from the parallelism every point is characterized by the permanent size F. Moreover, in each point the size F simultaneously characterizes the radius of sphere and the negative radius of tractrix. In other words, all spheres (both with positive and with negative curvature) have permanent size in every point.

3.3. Dynamics of the informative reality system

According to the informative reality theory suggested by the author, the Informative Reality System works as follows.

The expounded theory describes the fact of the presence of magnetic properties in revolving water molecules. The rotation of positive protons creates a solenoid as shown in figure 1.

The rotational motions of water molecules which move on the pushing space acquire a high speed. When the molecules have crossed the inversion layer, Figure 2, the transformation of quantity takes place to quality. Centrifugal forces begin to exceed the valence bond and the H_2O molecule is torn into the H^+ proton and the OH^- group. As a result, the energy of the rotational motion of the water molecule is passed to the tangential wind.

Straight up motion of the H_2O molecules interpenetrates to tangential wind of the H^+ and OH^- ions because of the interpenetration of opposites.

The property of permanent size will be taken as a base for description of the Informative Reality System.

Thus, informative reality operates on chaotic motion of molecules and organizes it in one direction. Informative reality organizes the process that carries out the action of the third law of materialistic dialectics: negation of the negation. The Informative Reality System negates the heat death and gives the way of matter development.

The Informative Reality System allows understanding such phenomena as hurricanes/cyclones, birth of tails of comets, mechanism of pumping of astrophysical masers. The results of visible, infrared, and radio spectrometry are written done by informative reality theory. It describes in details polarization and the masses spectral dates. We will expose it in next chapters.

Chapter 4
The hurricanes power system

Among all natural catastrophes which take place on Earth, the greatest danger is caused by hurricanes/cyclones. Long-term studies conducted by the National Hurricane Center, Colorado State University, National Oceanic, Atmospheric Administration and many other groups [14], allow us to learn deeply the spectral and thermodynamic materials. But as there is no adequate model, the frontal attack on the hurricanes, adopted by "Storm Fury" and conducted for twenty years – from 1963 to 1983 – ended in a failure.

4.1. Problems of the thermodynamics model

Extensive research allowed us to investigate the causes and births of hurricanes zones, their height and ways of movements, to describe their structure and dynamic. These "pricks" for a whirlwind energy which is about 10^{17} J with a frontal closeness about 100 J/cm^2 were rather weak. Atlantic basin seasonal hurricane investigations allowed us to build a capable working model of prognoses of hurricanes appearances.

As a result of wide research of birth, development, and fading of hurricanes, the volume of various measuring of physical parameters is accumulated. Repetition of data confirms their high truth. These results show that in hurricanes there are such processes which thermodynamics theory is not able to describe. We will specify them. Warm, dry, and transparent air in the eye of hurricane comes down from troposphere, becoming moist and cold near the surface of the ocean [47]. Moist and cold air rises from the surface of the ocean along the eye wall of a hurricane with an increasing speed creating here a tangential force to wind [33,34].

Hurricanes make highly charged electric clouds which can burst by lightings. Hurricanes rotate in opposite directions in different hemispheres, and the Earth magnetic field is opposite in two hemispheres. During the most studies, Gulf of Mexico hurricanes tend to migrate dominantly northward in the Earth magnetic direction field [48]. If the hurricane forms a giant system of magnets, this behavior seems to be logical.

4.2. Energy source of hurricanes

The studies of hurricanes by a spectral technique (polarimetry, infrared and mass spectroscopy) have showed that the ruling material force make the warm water molecules become an energy source of hurricanes[18].

The informative reality theory describes the fact of the presence of magnetic properties in revolving water molecules. The rotation of positive protons creates a solenoid, as shown on figure 3. The informing reality process forms a giant system of magnets. Solenoids cluster along the lines of the magnetic field, thereby creating an anisotropic environment that changes the polarization of light, as shown on figure 5. It is possible to assert that the magnetic acceleration of the rotational motion of a water molecule creates powerful infrared radiation. It is known that this global atmospheric phenomenon is accompanied by an anomalous increase of intensity at the

21

frequency of 22.24 GHz. It is the frequency of the rotational spectrum of the water molecule. Meteorological satellites work in this spectral range, as shown on figure 6.

According to the informative reality theory suggested by the author, the rotational motion of water molecules which have crossed the inversion layer in the eye of hurricane acquires a high speed.

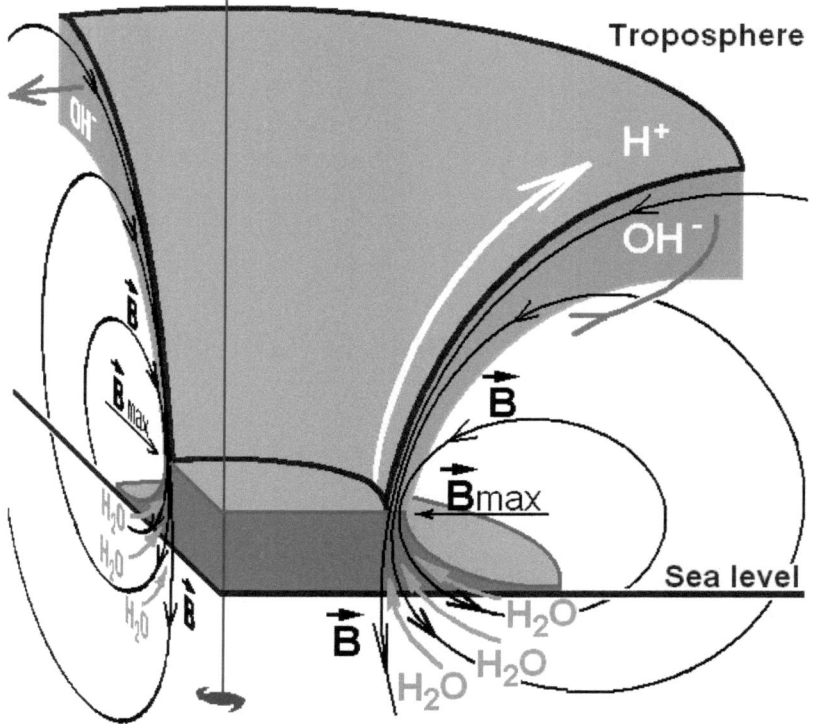

Figure 3. The hurricane/cyclone power system [21].

Centrifugal forces begin to exceed the valence bond and the H_2O molecule is torn into the H^+ proton and the OH^- group. As a result, the energy of the rotational motion of the water molecule is passed to the tangential wind of hurricane. Protons are moved by the eye wall of hurricane to the troposphere and form a proton cloud there. This result is based on the mass spectroscopy detection technique, which was implemented on the shuttle Discovery in 1998, as shown on the figure 7.

The suggested informative reality theory exposes the nature of the self-formation of a hurricane, specifies its forces, and adequately describes its parameters.

4.3. Spectral data

The expounded informative reality describes spectral data as well as thermodynamic materials [18]. We will show this further.

Polarimetry. The rotation of positive protons creates a solenoid. Solenoids cluster along magnetic field lines, thereby creating an anisotropic environment that changes the polarization of light. Figure 4 depicts polarization of the hurricane's environment by white bars. We can clearly see the growth of polarization around the eye wall. Polarization diminishes with the growth of distance from an eye wall.

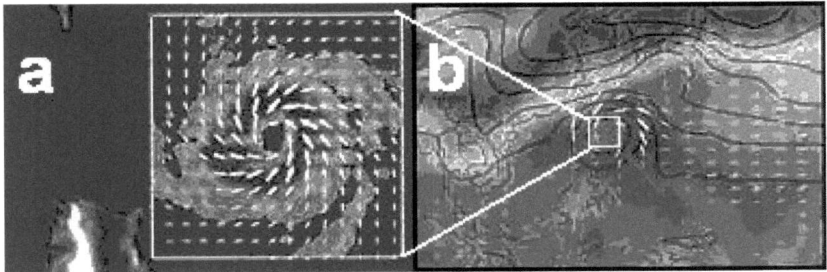

Figure 4. The solenoids cluster along magnetic field lines create an anisotropic environment and change the polarization of light [38].

Infrared radiation. It is possible to assert that the magnetic acceleration of the rotational motion of the water molecule creates powerful infrared radiation which is marked in figure 5. It is known that this global atmospheric phenomenon is accompanied by an anomalous increase of intensity at a frequency of 22.24 GHz. It is the frequency of the rotational spectrum of the water molecule. Meteorological satellites work in this spectral range.

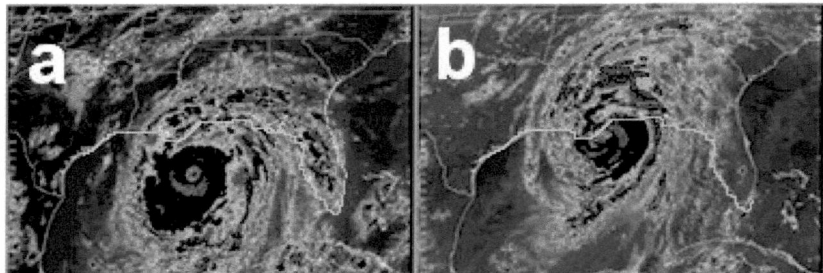

Figure 5. The rotational motion of the water molecule creates powerful infrared radiation [42].

Figure 5 demonstrates hurricane Katrina in infrared spectral range at the frequency of 22.24 GHz, when it moved above warm water in the Gulf of Mexico. It allows us to consider the following two scenarios. Figure 6a shows the infrared radiation produced by the rotational motion of water molecules. Twelve hours later the eye wall was disrupted by interaction with the land surface, where the water molecules are absent. As a result, the infrared radiation above the land surface decreased significantly in magnitude, as shown in Figure 5b.

Mass spectroscopy. As shown in figure 6, protons move by the eye wall of hurricane to the troposphere and form proton clouds there.

This result is based on the mass spectroscopy detection technique, which was implemented on the shuttle Discovery in 1998 [24].

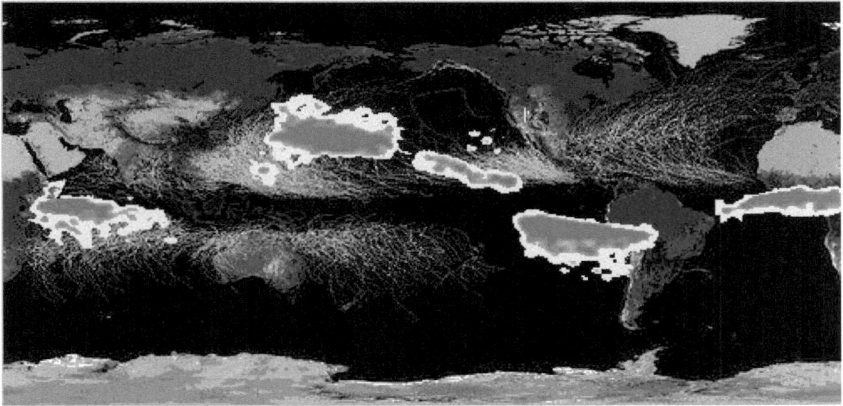

Figure 6. The map of way of hurricanes is given by points [40]. The proton's clouds [24].

The location of protons clouds correlates strongly with the ways of hurricanes, which are given by points [40].

4.4. Speed of maximum wind computations

The informative reality theory allows describing not only spectral materials of hurricanes but also its basic dynamic parameters.

Based on the principle that equal potential surface is a portion of the usual sphere of radius F, which is limited by the radius R, corporal corner filling of an equal potential surface is given by:

$$\Omega \, (R) \; = \; 2 \, \pi \, (1 \; - \; \frac{\sqrt{F^{\,2} - R^{\,2}}}{F}). \tag{17}$$

On the grounds that all magnetic field lines remain within the limits of this corporal corner, the dependence of magnetic flux density from R can be defined as:

$$\vec{B} \, (R) \; = \; 2 \, \pi \, \vec{B}_0 \big/ \Omega \, (R) . \tag{18}$$

We want to find solenoid kinetic energy, which it acquires passing a part of information reality space field. If the potential of solenoid in the point of $R = F$ is defined as equal to zero, kinetic energy of solenoid becomes equal:

$$E \, (R) \; = \; 2 \, \pi p_{\,m} \, B \, (R). \tag{19}$$

Taking the kinetic energy through the water molecules mass m and its speed $V(R)$ on the equal potential surface of R, we can find that the resulting equation (19) in relation to speed, is [19]:

$$V \, (R) \; = \; 2 \, \sqrt{\frac{\pi p_{\,m} \, B \, (R)}{m}}. \tag{20}$$

We will explain the dependence of the speed of maximum wind *Vmax* on the radius of maximum wind *Rmax*. The maximum wind *Vmax* can be defined from the equation (20) for maximum of the magnetic flux density vector *Bmax*. The radius of maximum wind *Rmax* will be found from Lorenz force (3), which operates on a charge in the magnetic field:

$$B_{\text{max}} \; = \; \frac{m_{\,p} V_{\,d}}{q R_{\text{max}}}. \tag{21}$$

where m_p is mass of proton, q is charge of proton, V_d is speed of OH⁻ group at dissociation, Bmax is magnetic flux density vector in a maximum.

We will get maximal magnetic flux density vector *of Bmax* as defined in equation (21) and applied in (20), we will *the following:*

$$V(R) = 2\sqrt{\frac{\pi p_m V_d}{qR_{max}}} + V_d \tag{22}$$

For verification of equation (22) the experimental data, in particular, the maximum wind speed *Vmax*, which conducts a hurricane to swirling.

The maximum wind speed is a function only of the radius of maximum wind *Rmax* [16]:

$$V_{max} = \frac{102}{\sqrt{R_{max}}} + 25 \ (km \ / \ h). \tag{23}$$

The curve on figure 7 represents the result of computation by the equation (23). The experimental data were taken from the paper by Shea Dennis J. and Gray William M. [35].

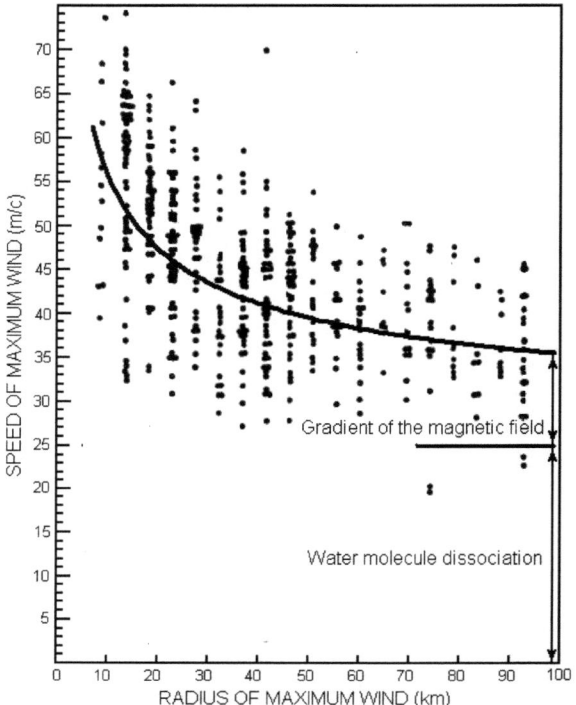

Figure 7. The speed of maximum wind computations with radius of maximum wind for lower troposphere data [16]. The equation curve is indicated by the line.

It is obvious from figure 8 that the results of measuring and computations on the equation coincide well.

Chapter 5
The comets power system

Being the least massive and the most intriguing objects, comets are a subject of astronomical examination and investigation. What is more, they are the object of ever increasing interest and physical study. A significant part of this new effort stems from the fact that comets act as natural probes of the interplanetary plasma or solar wind, and potentially contain a large amount of information which is difficult to obtain directly through space probes (such as properties away from the plane of the ecliptic).

5.1. Comets

Comets are small bodies of frozen volatiles (it is ordinary H_2O) embedded in a crusty silicate filler. According to modern presentations, comets are "small bodies" of the sunny system with continuously renovating atmospheres. They orbit the sun in an eccentric orbit. As comets approach the sun, they volatile to the atmosphere around the comet called the coma, and later a tail, around the nucleus.

In addition, comets provide a valuable supplement to our natural plasma physics laboratory already found in the corona and solar wind. The magneto hydrodynamic model seems be required to give a proper explanation of comet tails.

All comets are composed of a nucleus, coma, and tail. The comet nucleus is presumed to be a solid body with a radius in the range 1 to 20 km; masses of comet nuclei are very uncertain, but values of $\sim 10^{18}$ to 10^{21} g are found in the literature. The nucleus is thought to correspond to Whipple's [43,44]. "Ice-conglomerate" model of the nucleus consists of relatively complex parent compounds (possibly H_2O, NH_3, CH_4, CO_2, NH_2, etc.) which sublimate in vacuum at temperatures of a few hundred °K and when dissociated provide the daughter molecules observed in spectra of comets. Meteoric material of a wide size range is interspersed throughout the ices.

Calculations for a model nucleus (Brandt) [4] indicate a surface temperature in the range of 150-250°K. A superficial layer mainly consists of refractory components. Finally, the total mass loss a large comet during a perihelion passage has been estimated at 1 percent of the total mass; this mass must come from the nucleus.

The nucleus is the source of comet atmosphere. When approaching to the perihelion (to the sun) it is warmed up stronger, evaporation increases, the comet becomes overgrown with an atmosphere. The coma and tail appear.

The coma is composed of meteoric dust and neutral molecules in an essentially spherical volume centered on the nucleus with velocities about 0.5 km/sec [3] estimated from concentric expanding rings or halos from comet Halley. The coma can be detected out to distances of 10^5 to 10^6 km from the nucleus.

A comet's tail and coma are illuminated by the Sun and may become visible from Earth when a comet passes through the inner Solar System, the dust reflecting sunlight directly. Most comets are too faint to be visible without the aid of a telescope, but a few each decade become bright enough to be visible to the naked eye.

The streams of dust and gas thus released form a huge, extremely tenuous atmosphere around the comet called the coma, and the force exerted on the coma by the Sun's radiation pressure and solar wind form an enormous tail, which points away from the Sun.

The streams of dust and gas each form their own distinct tail, pointing in slightly different directions. The tail of dust is left behind in the comet's orbit in such a manner that it often forms a curved tail, only when it seems that it is directed towards the Sun. At the same time, the ion tail made of gases, always points along the streamlines of the solar wind.

As it turned out during the last years, interpretation of large complex effects in the comets caused by sunny corpuscular streams or sunny wind comes across considerable obstacles [29].

5.2. Questions posed to physics of comets:

Question 1. What kind of force does cause dissociation of molecules in the tails of comets?

As we can see, a large number of comet molecules are free radicals. It is assumed that they arise up as a result of dissociation and ionizing of neutral molecules, such as H_2O, CH_4, $NH_3(CN)_2$ [37,8].

Very often in immediate proximity to a nucleus spectral there are the ions of OH^- and CO^+. Usually the area of their distribution does not exceed 5000 km (the center is in the nucleus). As numerous spectral researches have showed [36], these tails consist mainly of the ionized molecules, it is plasma.

The reasons of appearance of ions are not clear in the tails of comets and comet atmosphere - none of credible mechanisms can provide the measured ionization. For example, the Sun's radiation must be extraordinarily powerful so that it is possible to explain the ion tail phenomenon [22]. The power of radiation must be 100-1000 times larger than corpuscular particles from the sun.

Such situation arises up because a comet atmosphere and corpuscular streams are extraordinarily discharged; the free run of proton of stream to the collision with a comet molecule is 100-1000 times longer than the linear cross runner of middle brightness comet coma.

Question 2. What is the mechanism of a magnetic-field birth in the tail of comet?

In 1957 H. Alfven [1] marked the important role of the magnetic fields in the frozen streams. His work was a push to the development of a magnetic hydrodynamic method in physics of comets. S. B. Pikelbner stated [31], that co-operating of stream with the atmosphere of comet can be realized by means of magnetic-field. Cooperation also takes place without collisions.

Magnetic hydrodynamics helped overcome many difficulties, yet most problems are distant from the quantitative (sometimes even qualitative) understanding.

Question 3. It is not clear, for what reason the particles suddenly begin to move one-way from the nucleus of comet in external space.

Whitney [45] calculated middle mass of the thrown out matter for the row of comets to prove the hypothesis that hard particles are thrown out measuring 10^{-7} m. He got mass of a $2*10^{10}$ g that at speed of 1 km/c gives 10^{20} erg. O. V. Dobrovolsky [9] did an analogical calculation to test the hypothesis that halos consists of a gas cloud. He got the analogical result. On the whole, energy freed corresponds to energy of sunny radiation that will be got by the nucleus of comet.

As we can see, the energy balance is good enough. But unfortunately, there is a matter with a mechanical motion of molecules in the tails of comet nuclei. Here a mechanical theory is powerless to explain the variety of the phenomena under investigation even in basic lines. Under the action of what forces are water molecules with high speed pushed from the comet nucleus?

The mechanical theory can not describe, for example, such phenomena, as enormous accelerations in the tails, and transversal motion of substance in a tail, perpendicular to the radius vector.

All these phenomena are insoluble within the framework of both mechanical and magnetic hydrodynamics.

The informative reality gives the key to understanding all the variety of these phenomena.

5.3. The Informative Reality System in comets

As a result of long-term research of birth, life, and fading of comet's tails and comas, the volume of various measuring of physical parameters is accumulated. Repetition of data confirms their high truth. These results testify the following processes in comets.

The informative reality theory can help explain many facts about comets. In addition to many contributions to our understanding of the formation and evolution of comet coma, the informative reality theory will also provide an excellent opportunity to learn about the basic physics of ions in a cometary environment.

This paper, strictly speaking, not only concerns ion tails; it also raises the possibility for discussing them with giving a reasonable idea of a total comet model.

The studies of comet tails and comas by a spectral technique (polarimetry, microwave, infrared, and visible spectroscopy) shows that magnetic, dynamic, and electric forces of the warm water molecules become energy source of comet atmosphere from easily evaporating ice of comet nucleus.

In connection with the low temperature of icy nucleus of comet the question becomes urgent - what minimum temperature excites the rotatory motion in the water molecules?

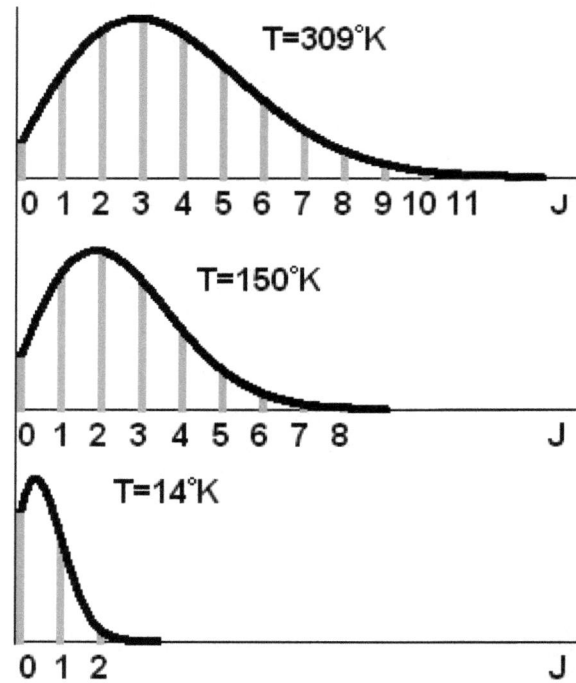

Figure 8. The rotational population of the water molecules at different temperatures.

The figure 8 presents the populated rotatory levels of H_2O with In=9,52 cm^{-1} for different temperatures.

34

As it is obvious from the figure 8 at the temperature 14°K the half of water molecules has rotatory motion.

At the temperature 150°K a maximal amount of molecules is at second rotatory level, and at the temperature 309°K a maximal amount of molecules is at third rotatory level. Thus, as we can see, the informative reality process can work at a temperature higher than 14°K. Calculations for the model nucleus (Brandt) [21] indicate the surface temperature in the range 150-250°K.

The expounded in the chapter 1 theory describes the fact of the presence of magnetic properties in revolving water molecules. The rotation of positive protons creates the solenoid as shown in figure 1. Solenoids cluster along the lines of the magnetic field, thereby creating an anisotropic environment that changes the polarization of light. It is possible to assert that the magnetic acceleration of the rotational motion of a water molecule creates powerful micro wave radiation. It is known that this global astrophysical phenomenon is accompanied by an anomalous increase of intensity at a frequency of 1665 MHz. It is the frequency of the rotational spectrum of the water molecule. Radio-telescopes work in this spectral range.

Figure 9 illustrates the informative reality model of the comet in which the water molecules have pumped in the tail when the jet formed.

Figure 9 also shows the comet body which consists of frozen volatiles (it is ordinary H_2O) embedded in crusty silicate filler. The filler has holes.

Approaching to the sun it starts to warm up stronger, evaporation increases, and water molecules begin to take off from the nucleus through the holes in outside. As a result, the Informative Reality System appears.

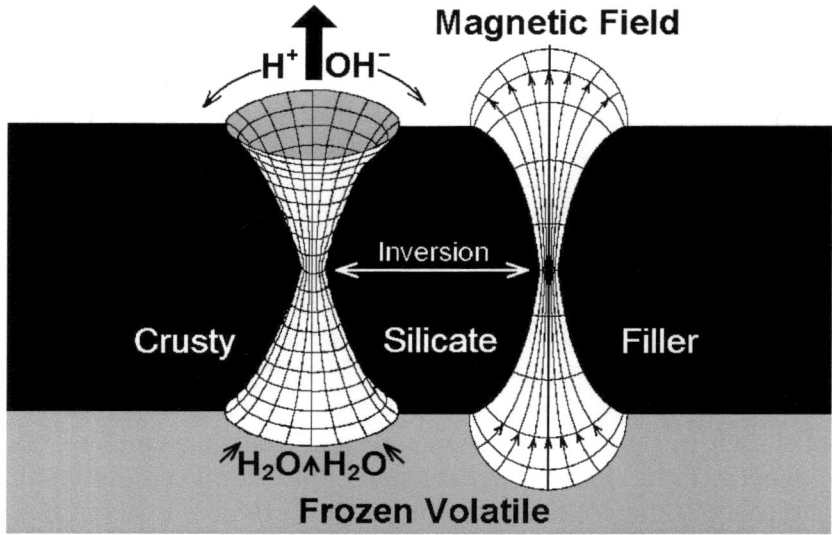

Figure 9. The comet's Informative Reality System.

According to the informative reality theory suggested by the author, the rotational motion of water molecules which have crossed the inversion layer acquires a high speed. Centrifugal forces begin to exceed the valence bond and the H_2O molecule is torn into the H^+ proton and the OH^- group.

5.4. Tow ions velocities

We can estimate the speed of water molecule rotation that breaks valency connection. Because energy of H^+ -OH^- valency connection is equal to $W=8,2*10^{-19}$J, we will find the protons velocity limit V from the equalization:

$$W = \frac{mV^2}{2},\qquad(24)$$

where m is the total mass of two protons.

The protons speed limit V is easy to use for simple calculations.

We compute from equation (24) a velocity limit V_H=28 km/c. It is the tangential velocity of H^+ proton. From equality of impulses of H^+ proton and OH^- group we can find speed of OH^- group V_{OH}=5.3 km/c.

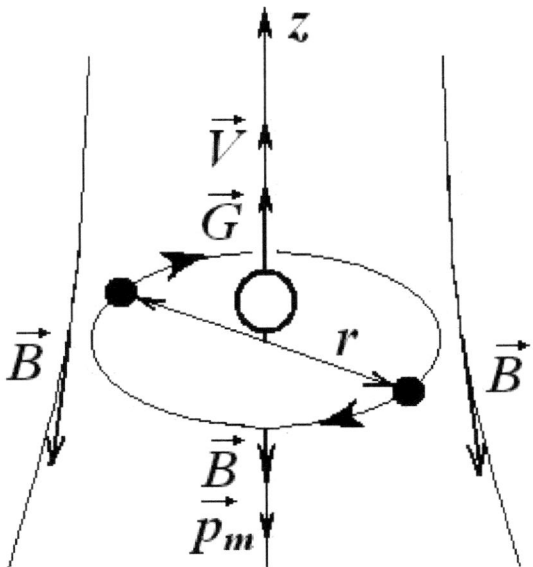

Figure 10. The rotational motion of the water molecule.

As a result, the energy of the rotational motion of the water molecule is passed to the tangential motion in the comet halo. Ions move from the nucleus to the comet tail and form the ions cloud there.

The suggested informative reality theory exposes the nature of the self-formation of comet coma and tail, specifies its forces, and adequately describes its parameters.

Chapter 6
The astrophysical masers power system.

Night sky is filled with enormous amount of various objects; such as stars, galaxies, and clouds. We can see them due to their ability to radiate light. A radiation takes place by law of a heated body. A body with the fixed temperature has the fixed color. For example, a stove (a temperature of about 400 °C) shines the red color and the glow lamp (a temperature of about 3000 °C) shines the white color. These bodies shed light in the state of thermodynamics equilibrium.

However, there are objects in space, which do not obey this rule, they are the astrophysical masers. The astrophysical masers are naturally occurring sources of stimulated spectral line emission, in the microwave portion of the electromagnetic spectrum. They are cold space clouds in which the rotation of molecules is strongly excited abnormally.

6.1. Astrophysical masers

The name MASER comes from the capitalized initials of the following words: Microwave Amplification through Stimulated Emission of Radiation. Microwave radiation is emitted when a transition occurs between rotational levels of a molecule with inverted populations.

The emission from a maser is stimulated and monochromatic, having the frequency corresponding to the energy difference between two quantum-mechanical energy levels which have a non-thermal population distribution.

Historical background. In 1965 an unexpected discovery was made by Weaver et al. [41]: emission lines in space, of unknown origin, at a frequency of 1665 MHz. It is well known that molecules could not exist in space, and so the emission was at first attributed to an unknown interstellar species named Mysterium, but the emission was soon identified as line emission from OH molecules in compact sources within molecular clouds [10]. More discoveries followed, with H_2O emission in 1969 [5], all coming from within molecular clouds. These were termed "masers", as from their narrow line-widths and high effective temperatures it became clear that these sources were amplifying microwave radiation. Masers were also discovered in external galaxies in 1973 [46].

Another unexpected discovery was made in 1982 with the discovery of emission from an extra-galactic source with an unrivalled luminosity about 106 times larger than any previous source [2]. This was termed a megamaser because of its great luminosity; many more megamasers have since been discovered. However, the question of origin of the molecular clouds remains open until now.

Line narrowing. The irregularly shaped maser cloud becomes greatly distorted by exponential gain. Part of the cloud that is slightly longer then the rest will appear much brighter, and so maser spots are typically much smaller then their parent clouds. The greatest part of the radiation will emerge along this line of greatest path length in a "beam"; this is termed beaming.

Exponential gain also amplifies the centre of the line shape (Gaussian etc.) more than the edges or wings. This results in an emission line shape that is much taller but not much wider. Consequently, it is necessary to talk about a high-velocity molecular stream forced under the pressure out a small-diameter nozzle. But where is this nozzle, and what force starts it going?

High brightness. The brightness temperature of the maser is the temperature a black body would have producing the same emission brightness at the wavelength of the maser. That is to say, if an object had a temperature of about 10^9K, it would produce as much 1665 MHz radiation as a strong interstellar OH maser. As a matter of fact, at 10^9K the OH molecule would dissociate (kT is greater than the bond energy), so the brightness temperature is not directly indicative of the kinetic temperature of the maser gas.

Polarisation. An important aspect of maser study is polarization of the emission. Astronomical masers are often very highly polarized, sometimes 100% (in the case of some OH masers) in a circular fashion. This polarization is due to some combination of the Zeeman effect, magnetic beaming of the maser radiation, and anisotropic.

Unlike terrestrial lasers and masers for which the excitation mechanism is known and engineered, the reverse is true for astrophysical masers. In general, astrophysical masers are discovered empirically then studied further in order to develop plausible suggestions about possible pumping schemes. Quantification of the transverse size, spatial and temporal variations, and polarization state are all useful in the looking for a pump theory.

What does carry out pumping of inverted populations?

Maser oscillation at these wavelengths can be achieved by pumping with reasonable amounts of incoherent pump light.

6.2. Terrestrial laboratory maser

The major differences between laboratory masers and astronomical masers should be mentioned.

The resonant cavity. In laboratory masers, the maser region is typically ~1 m long. In order to have significant gain, the radiation is bounced between tow mirrors, one of which is semitransparent to allow the amplified radiation to escape. In contrast, astronomical masers are single-pass, broadband, and have liner dimensions of a few AU (about 150 million km) or larger. Because the radiation transfer in astronomical masers is relatively simple, the main theoretical effort has concentrated on inventing efficient pumping mechanisms.

Plenty of researches show absence of a resonant cavity at astronomical masers that compels to suppose fundamentally other mechanism of the powerful micro wave radiation.

The pumping mechanism. The mechanism that leads to inversion, which is caused by the cycling of molecules through energy levels, is called the pump. The pump process is initiated by excitations from the ground state, due to either external radiation or collisions. The inversion results from a combination of various factors, specific to the particular maser molecule.

Before the construction of pump model of the astrophysical maser would profit to consider work of the laboratory maser that is fully studied in all details.

In laboratory masers, the maser region is typically ~ 1 m long. In order to have significant gain, the radiation is bounced between tow mirrors, one of which is semitransparent to allow the amplified radiation to escape.

Spatial coherence allows the maser to be focused to a tight spot. Laser beams can be focused to very tiny spots, achieving a very high irradiance, or they can be launched into beams of very low divergence in order to concentrate their power at a large distance.

In laboratory masers, the inverse process, induced or stimulated emission, is the essence of the maser phenomenon. In it, a downward transition is induced by an incoming photon with a matched frequency. To conserve energy and momentum, the transition is accompanied by the emission of another photon whose properties are identical to those of the initial parent photon.

The inversion of populated levels is got by means of non-coherent light of pumping. The incoherent light is absorbed by the active medium so that the molecules are pumped to the upper maser level; it is level 2 on Figure 11.

- Optical Pumping: Atoms are excited to higher energy levels through absorption of photons. Incoherent lamps or lasers can be used.

- Electrical Pumping: Electrical discharge (in gases) or current (in semiconductors) excited the atoms.

The two processes above are the most common. Less used, yet possible other mechanisms are:

- X-Ray pumping;
- e-beam pumping;
- Chemical pumping;
- Gas dynamic pumping.

The major differences between the pumping of laboratory masers and astronomical masers are the following.

The laboratory masers are multi-pass, single-directed; the radiation is bounced between tow mirrors with region ~ 1 m long. In contrast, astronomical masers are single-pass, much directed; have liner dimensions of a few AU (about 150 million km) or larger, and have no mirrors.

In addition, wide researches showed absence of reasonable amounts of incoherent pump light that leads to inversion, which is caused by the cycling of molecules through energy levels. In astronomic maser the cyclic moving of molecule from one level to other is absent, and the emitted molecule does not get pumping of power level. The fact of absence of the reasonable amounts incoherent pump light is experimentally confirmed.

6.3. The astronomical maser pumping model

All totality of the astronomic maser properties results in a conclusion, that the pump of astronomic maser does not excite the molecules in the maser cloud. The excitation molecules process takes place elsewhere. By analogy with comets, we will name this place a nucleus. Thus, the pump of astronomic maser is in the nucleus. Maser molecules can be excited of necessary amount in the nucleus and pumped to the maser cloud. In maser cloud the molecules emit the radiation, pass to a minor level, and disperse in surrounding space. For this reason, the amount radiated by the maser cloud of quanta equals to the amount of pumped to the maser cloud molecules.

This process described in the previous part of the manuscript for comets. As a result, bright beam in maser is the tail of comet or halo, and spot is the coma or comet atmosphere.

In the process of pumping of astronomic maser none works of five mechanisms that are used for pumping of laboratory maser listed above. Here is not a non-coherent light pump.

Wide researches showed that the astronomic masers pumps have the informative reality nature described in the first part, we will expose it.

The maser power system consists of three zones. Zone a) the rotational H_2O magnet dipoles are pulled in by gradient force. Zone b) the centrifugal force breaks the rotational H_2O molecules on H^+ and OH^- ions. Zone c) the dielectric plasma pushes the ions from magnetic field into maser cloud.

Figure 10 shows the chart of maser pump that consists of the following elements.

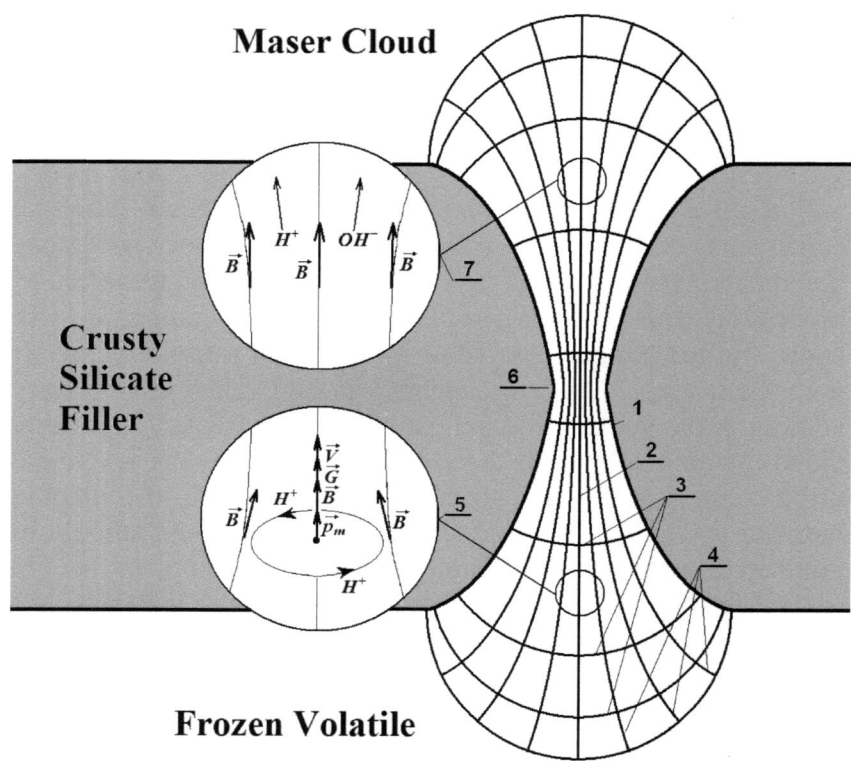

Figure 10. The Informative Reality System carries out pumping of inverted populations [23].

1) The tractric nozzle. 2) The axis of symmetry. 3) The equal potential surfaces, which are spheres with the constant radius F. 4) The magnetic field lines, which are tractrices of the constant negative radius F. 5) Zone of involvement of magnetic dipoles with four parallel vectors: force $\vec{G}$, speed $\vec{V}$, solenoid $\vec{p}_m$, and magnetic induction $\vec{B}$. 6) The inversion layer where the magnetic field achieves his maximum and revolving molecules are torn into ions. 7) The magnetic field,

which pushes the diamagnetic ions (the H^+ protons and the OH^- groups) to the maser cloud.

Now we will consider all these elements of the maser power system separately.

The molecule of water, as it is known, consists of oxygen atom and two positively charged atoms of hydrogen (protons), the rotation of protons creates the magnet. According to the informative reality theory suggested in chapter 2, co-operation of the informative reality with water molecules solenoids results in the turn of the solenoid vector in direction of magnetic flux density vector, and, as a result, strengthens the magnetic field. The rotational motion of water molecules acquires a high speed and, accordingly, the inversion of rotation levels.

Centrifugal forces begin to exceed the valence bond and the H_2O molecule is torn into the H^+ proton and the OH^- group. As a result, the energy of the rotational motion of the water molecule is passed to the tangential moving of ions. They are moved along a line of greatest path length in a "beam" and form a maser cloud.

An electric motor uses energy of magnetic field to produce a mechanical work; it is very typical through the interaction with ions of H^+ and OH^-. Ions are pushed from the magnetic field to the maser cloud, it is executed in accordance with the plasma theory.

Figure 11 shows both the chart of laboratory laser pumping, the picture on the left and astrophysical maser pumping, the picture on the right.

The inversion of populated of levels in the laboratory laser pumping is got by means of visible light of pumping. The light is absorbed by the active medium so that the molecules are pumped from lower laser level to the upper laser level; it is level 2 on Figure 11. The laser transition takes place from the upper level 2 on the lower level 1.

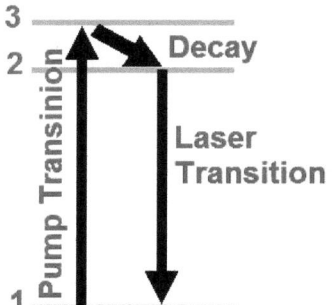
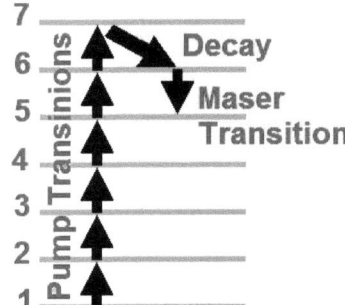

Figure 11. The pumping schemes of the laboratory laser and the astrophysical maser.

As it follows from the informative reality theory, pumping of astrophysical maser takes place consistently a few transitions. Pumping takes place the not photons of visible light, which is not present in a dark cavity, but thermal photons. Coming from the calculations brought around to Figure 8, thermal photons appear at the temperature of 14°K and higher. Pumping top level comes true for a few transitions. The maser transition comes not on the lower level 1. Thus, with maser on the frequency of 22.24 GHz of the rotational spectrum of the water molecule, the transition comes from the 6_{16} to 5_{23} level.

The theory of informative reality and interpretation presented here suggest strongly that the inverse of the maser clouds molecules obeys the conditioned model in which the heat from the frozen volatile nucleus grows into rotation motion of maser molecules.

The informative reality theory describes and, consequently, allows us to explain all effects listed above.

Chapter 7
Power of knowledge in economy

Modern economy is an energy expense. All of her is based on the booty of power mediums as coal, oil, to uranium and others like that and receipt of useful work due to their incineration.

The implementation of knowledge power in economy gives substantial results. For example, it improves functional possibilities and increases an efficiency of the system that prepares engine's air – fuel mixture.

7.1. The Informative Reality System for a car engine

The Informative Reality System for preparation of engine's air – fuel mixture was offered by authors [17]. A device is technological; it allows optimizing work of transport vehicles and getting the economy of fuel. He is set between the choke vale and the intake valve.

The device is based on the physical phenomenon of water molecules dissociation, when molecules move along informative reality. As it is well known, air contains 0.2-0.5% water pairs. Air follows in engine at a necessary amount.

The Informative Reality System for the car engine is shown on figure 12.

Figure 12. The Informative Reality System for the car engine.

The scheme of Informative Reality System is given on figure 13. The Informative Reality System creates pulling space and pushing space. The inversion layer divides them.

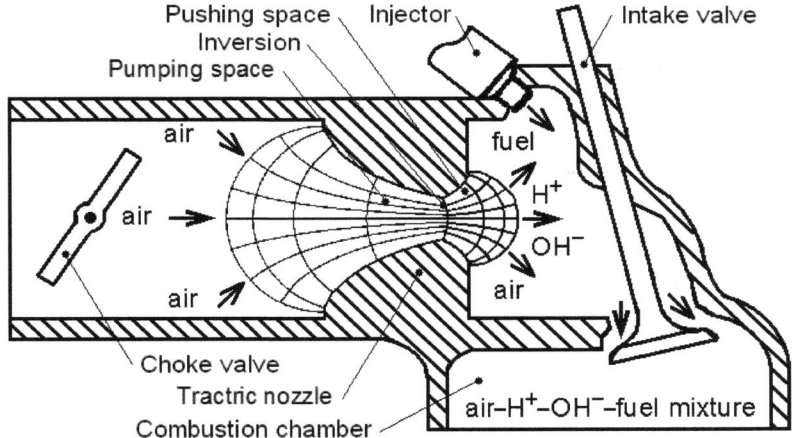

Figure. 13. The scheme shows transformation of air and fuel into the air – H$^+$ – OH$^-$ – fuel mixture by the Informative Reality System.

In accordance with the informative reality theory, the rotational motion of water molecules which have crossed the inversion level acquires a high speed. Centrifugal forces begin to exceed the valence bond and the H_2O molecule is torn into the H$^+$ proton and the OH$^-$ group. The presence of ions promotes engine power considerably.

The usage of the Informative Reality System in combustion engines leads to a driving power increase up to 30% - 40% and petrol consumption reduces by 25% - 30%.

Researches [23] showed that the cost of re-equipment an engine by informative reality system would be recompensed for 2 months.

As been shown on the 8-th International Green Energy Conference [15], a car with the Informative Reality System diminishes an impact on the environment in terms of air quality, greenhouse gases, ozone depletion, water quality, use of natural resources. Simultaneously there is a multiple decrease in engine vibrations.

Components (%)	Petrol's engines	Petrol's engines with IR System	Diesel's engines	Diesel's engines with IR System
Nitrogen	74-77	76 - 78	76 - 78	76 - 78
Oxygen	0,3 - 5	6 - 12	2 - 8	7 - 16
Dioxide of carbon	5 - 12	2 - 6	1 - 10	0,5 - 3
Oxide of carbon	1 - 10	0,02 - 0,8	0,01 - 0,5	0,001 - 0,01
Water vapor	3 - 5,5	0,2 - 0,8	0,5 - 4	0,1 - 0,5
Oxides of nitrogen	0 - 0,8	0 - 0,8	0,001 - 0,4	0,001 - 0,4
Hydrocarbons	0,2 - 3	0,02 - 0,4	0,01 - 0,1	0,001 - 0,04
Aldehydes	0 - 0,2	0 - 0,001	0 - 0,002	0 - 0,001

The combustion engine functioning with such an Informative Reality System gets improved firing process and exhaust causes minimal air pollution, what you can see in the table [22].

Therefore, the usage of Informative Reality System in combustion engines doubles an amount of oxygen in exhausts, at the time content of harmful substances reduces significantly. An important fact is that exhausts contain no carbon (soot).

Informative Reality System provides conditions, under which an engine certified as "Euro-3" according to the European standards turns to the requirements of the engine certified as "Euro-5" that substantially diminishes contamination of atmosphere.

7.2. Intelligent Energy

All economies of the world spare enormous attention to the energy-savings. It is a strategic European initiative to stimulate on the topics of energy efficiency and renewable energy. It is part of the European Commission's Energy Efficiency Plan and is funded by the Intelligent Energy Europe (IEE) programme. IEE is a European Union funding programme to promote energy efficiency and renewable energy in Europe, what supports 280 international projects [25], 1400 Europe organizations that create the best terms for more intellectual use of energy in the future are attracted in that more than. Projects cover all industries of economy, including proceeding in energy, effective for energy building, industry, consumer wares, transport and support of countries that develop.

Here it should be noted that the hurricane/cyclones, comet, and astrophysical maser pump has been successfully realized in economy, as the Informative Reality System.

Informative reality works so that considerable part of energy comes from an environment, namely from air, earth, sea. It diminishes part of power mediums in a unit cost what substantially promotes the competitiveness of economy.

As be shown in the manuscript, the Informative Reality System works due to ruling forces of nature, in other words, due to the use of Intelligent Energy.

Conclusions

The material expounded in the manuscript confirms once again, that the laws of materialistic dialectics with identical force operate both in the economic and in the physical spheres.

The theory and interpretation presented here suggest strongly that the powers both in economy and in nature are divided into two classes: working class and ruling class. The ruling power obeys the following law: knowledge is power which rules the motion without implementation of any work.

Ruling power in economy, called "Professional and Managerial Class", is defined by the organizing function it plays in the reproduction.

Ruling power in nature, which always operates across direction of motion, exists and is generally known. It is Lorenz power. Power of knowledge (or Lorenz power) exists independently, out of human consciousness, creating informative reality.

Thus, on condition that we examine Lorenz power, as power that organizes motion in some set direction, many difficult phenomena of nature become clear. The informative reality gives force to the hurricanes/cyclones, comets, and astrophysical masers. The informative reality theory allows describing qualitatively and quantitatively the enormous amount of measurement of all these various phenomena of nature. Repetition of data confirms their high truth.

The suggested approach allows us to improve the experimental technique for investigation of nature and gives unlimited possibilities for managing by reproduction.

As it has been shown in the manuscript, informative reality has wide distribution in nature. It can often be found in the atmosphere of Earth, in the sunny system, in galaxies, and in intergalactic space. Everywhere nature carries out the ruling action.

All things considered, on the question of Engels to Darwin: does nature choose?

The power of knowledge allows us to give the negative answer:

No, nature does not choose, - on the contrary, nature rules.

References

1. H. Alfven,Tellus **9**, 92 (1957) .

2. Baan W.A., Wood P.A.D., Haschick A.D. 1982 Astrophys. J. 260 L49-52

3. N.T. Bobrovnikoff, Lick. Obs. Publ. **17**, 309 (1931) .

4. Brandt, J. C. The Physics of Comet Tails. Annual Review of Astronomy and Astrophysics, vol. 6,(1968) p.267-286.

5. Cheung A.C., Rank D.M., Townes C.H., Thornton D.D., Welch W.J., Crowther J.H. 1969 Nature 221 626-8

6. Engels F. Dialectics_of_Nature // http://www.marxists.org /archive/marx/works/download/ ngelsDialectics_of_Nature_part.pdf

7. Erik Olin Wright, Varieties of Marxist Conceptions of Class Structure//Politics & Society 9, no.3 (1980): 323-70

8.В.И.Чередниченко , Диссертация (Киев,1956).

9. Добровольский О.В., Астрон.ж. **31**, 324(1954).

10. Davis R.D., Rowson B., Booth R.S., Cooper A.J., Gent H., Adgie R.L., Crowther J.H. 1967 Nature 213 1109-10

11. Єрмошенко М.М. Інформація в системі виробничих відносин //Актуальні проблеми економіки. – 2007. – №10. – С. 66–73.

12. Gerhard Lenski, Power and Priviledge (N.Y.: McGraw Hill, 1966).

13. Jacob, J.R., Boyle's Atomism and the Restoration Assault on Pagan Naturalism, Social Studies of Science, Volume 8, Issue 2 , May 1978, 211-233.

14. Klotzbach Philip J., "On the Madden-Julian Oscillation-Atlantic Hurricane Relationship," Journal of Climate 23, 282–293 (2010).

15. Kucherov O. The Nature of Hurricane Green Energy // Proceedings of the 8-th International Green Energy Conference, Monograph, NAU, Kyiv-2013. — P.317-320.

16. Kucherov O. Power of knoweledge, which maneges direction of motion in economy and nature // Jornal of Qafqaz University. Mathematics and Computer Science. №1, 2013, p. 24-30.

17. Kucherov O.P., Pazdriy Y.E., Fuel Feed Device of an internal combustion engine. Patent of Ukraine № 90406 (2008).

18. Kucherov O. P. and Pazdriy Y. E. Advanced hurricane studies by a spectral detection technique: polarimetry, infrared and mass spectroscopy// NATO ASI "Spectral Detection Techcique (Polarimetry) and Remote Sensing" Kyiv, Ukraine, 12-25 September 2010.

19. Kucherov O.P., Pazdriy Y.E., Information Substance Decreasing Entropy of a Complex System // Actual Problems of Economics, Scientific Economics Journal. — 2010. — №9. — P.300-304.

20. Kucherov O.P., Pazdriy Y.E., Informative Reality, as a Mean of Hurricanes Managment // Actual Problems of Economics, Scientific Economics Journal. — 2011. — №2. — P.214-221.

21. Kucherov O.P., Pazdriy Y.E., Edvanced Hurricane Studies by Spectral Detection Technique: Polarimetry, Infrared and Mass Spectroscopy // Astronomical School's Report. 2011.— №1-2. — P.48-55.

22. Kucherov O.P., Pazdriy Y.E., Use of Information reality in the Ecological Management System // Actual Problems of Economics, Scientific Economics Journal. — 2012. — №1/2. — P.129-136.

23. Kucherov O.P., Pazdriy Y.E., Economic evaluation of the Informative Reality Influencing on Energy Consumption in the

Ukraine // Actual Problems of Economics, Scientific Economics Journal. –2011. – №7. –С. 307–313.

24. Luchkov B., "Hurricanes are eternal problem," Science and life 3, (2006).

25. Lambert P. News from the field. News review of the Intelligent Energy - Europe programme, 2007.-N1 - April .- p. 6

26. Ландау, Л. Д., Лифшиц, Е. М. Статистическая физика. Часть 1. — Издание 3-е, дополненное. — М.: Наука, 1976. — 584 с. — («Теоретическая физика», том V), §8.

27. Luchkov B., "Hurricanes are eternal problem," Science and life 3, (2006).

28. Marx Karl. "The German Ideology". Literary Theory: An Anthology. 2nd ed. Oxford: Blackwell, 1998. 658. Print

29. Марочник Л.С.. Магнитогидродинамические явления в кометах и связь их с геоактивными потоками. Успехи физических наук. Т.LXXXII, вып. 2, 1964, с.221-252.

30. Макаров А.М., Лунева Л.А., Основы электромагнетизма, Физика в техническом университете, Т. 3, Московский университет им. Баумана. (2002).

31. Пикельнер С.Б., Основы космической электродинамики, М., Физматгиз, 1961.

32. Ralf Dahrendorf, Class and Class Conflict in Industrial Society (PaloAlto: Stanford University Press, 1959);

33. Shapiro, L. J., Willoughby H. E., "The response of balanced hurricanes to local sources of heat and momentum," J. Atmos.Sci. 39, 378–394 (1982)

34. Schubert, W. H., and Hack J. J., "Inertial stability and tropical cyclone development," J. Atmos. Sci. 39, 1687–1697 (1982).

35. Shea Dennis J., Gray Willam M. "The Hurricane's Inner Core Region," Journal of the Atmospheric Sciences 30, 1544–1564 (1973).

36. Swings P., Haser H,. Atlas of Repr. Cometary Spectra, 1956.

37. Станюкович К.П. Неустановившиеся движения сплошной среды. М. Гостехиздат,1955.

38. Thomas R. Knutson, "Global Warming and Hurricanes, an Overview of Current Research Results," Climate Dynamics 15 (1999)

39. Thomson, W. (1851) "On the dynamical theory of heat; with numerical results deduced from Mr. Joule's equivalent of a thermal unit and M. Regnault's observations on steam" Math. and Phys. Papers vol. 1, pp 175–183

40. Tropical Cyclone. //http://ru.wikipedia.org/wiki/

41. Weaver H., Dieter N.H., Williams D.R.W., Lum W.T. 1965 Nature 208 29-31

42. Weber Jeff, "NCAR hurricane work reaches new intensity," Staff Notes, monthly, 9 (2005).

43. Whippie F ,Astrophys. **J.111** , 375 (1950).

44. Whipple F.,Astrophys. **J.113**, 464 (1951).

45. Whitney Ch, Astrophys. **J. 122**, 190 (1955) .

46. Whiteoak J.B., Gardner F.F. 1973 Astrophys. Lett. 15 211-5

47. Willoughby H. E., "Tropical Cyclone Eye Thermodynamics," Mon. Weather Rev. 126, 3053–3067 (1988).

48. Willoughby, H. E., "Temporal changes in the primary circulation in tropical cyclones," J. Atmos. Sci. 47, 242–264 (1990).

49. The Works of Francis Bacon, Lord Chancellor of England, 16 vols, ed. Basil Montagu (London: Pickering, 1825-1834), volume 14.

Printed by Books on Demand GmbH, Norderstedt / Germany